That Awesome Place Called Space

Your Illustrated Guide to what's Beyond the Sky!

Jon P Fox

DEDICATION

I just want to thank everyone involved in my childhood and early adulthood years that contributed to the knowledge that I bring forth on the subject of Space.

As a young mind I used to gaze out at the sky and just ponder what I am seeing. Twinkle twinkle little Star I used to say to myself as I look for a Star that is actually twinkling. It became a favorite subject of mine to learn about what is beyond the Earth and the science of what is out there. I painted a couple pictures of what I remembered a scene from outer Space to look like, but that was based on pictures from books and magazines I had paged through. This is because I had never actually been there, but something has been there and still is. To this day at the time of writing this book (2015), Scientists have sent out Space probe crafts that have been out there many years now and they send back awesome pictures.

And there is this giant Telescope that orbits the Earth that goes by the name Hubble. It can see across the universe for many light years away, not to mention the ground based telescopes that can also see to vast distances in Space. So I dedicate this book and attribute my making of it to those who explore the universe and bring back names and places that they found out there, even though they personally never left the Earth. And to those who have left the Earth's surface and hang around in the International Space Station that orbits the Earth. As men and women we are taking steps to one day explore Space beyond the Earth using manned Spacecraft that perhaps will travel fast enough to go long distances but also return from the mission. So thank you science community both here and internationally, and thank you teachers of mine growing up and into the age of the interfacing of devices and computers.

Copyright Information

CONTENTS

Jon P Fox

That Awesome Place Called Space

INTRODUCTION

Welcome and thank you for your interest in "That Awesome Place Called Space" – *Your Illustrated Guide to what's Beyond the Sky!*

When you look up from outside you see the sky. During the day it is blue and sometimes it has lots of clouds. You also see that there is this giant ball of fire up there that is so bright you hurt your eyes if you stare at it. At night you see a black sky and thousands of tiny little white dots everywhere, that are just way up there and they make you wonder. Your Mom or Dad or Teacher told you that they are Stars. This book is here to tell you more about what is up there. It is to tell you what is beyond that sky, so that you can get some answers and lets you see some examples of that frontier. **There exists an extremely vast universe we call Space, What is Out There?**

This book is made to help readers explore and appreciate outer Space. It aims to reward the curiosity of young minds by teaching them about what's out there -- from the study of *Astronomy*, the countless gigantic Stars, the magnificent Planets, and a lot of fantastic wonders of which can be seen among the great universe.

Now that you have this book in front of you it is time to open the pages and take a look at the wonders that we will discuss, I hope you will enjoy it and learn a lot from it! I hope you will come back to this book when you have questions, at least the fundamental ones as we don't have all the answers to our questions. ☺

Have   fun   exploring!

ACKNOWLEDGMENTS

It is thanks to National organizations worldwide that have compiled imagery and made it available to the public for educational purposes. Organizations and Agencies like NASA who are dedicated to the study of astronomical science as well as the physical universe, make it possible to bring visual aid to the subject matter and general explanations of Space, that are found in the chapters of this book. This book is not considered advanced in science but intended to be a fun learning experience to children and anyone else for the most part.

As the author I am very thankful for the availability of the resources that are at reach and am grateful that I can use these resources to help teach and get the point across to the reader of the text in this book. The images used here were obtained by way of public domain, free to use even commercially, license and are found to be available to anyone even without sponsorship. There are no images being used as advertisements or product endorsements of any kind but they are solely used to back up and be visual aid to the concepts and explanations of the book's text.

1 GENERAL SPACE INFORMATION

Let's start with this question: "How big is the outer Space?" This is one of the questions that made the Space beyond our atmosphere so awesome. Although we've been studying it for a very long time, it seems that every answer we find brings up even more questions than when we started.

But if you look at it differently, Space itself can actually be very, very small. In fact, Space is actually made up of what appears to be nothing. But that nothing is known now as *Dark Matter*. Aside from the particles, Stars, galaxies, remnant Nebulae, and other elementary bodies, outer Space is made up of *empty Space*, or empty *dark matter*.

Ah, but dark matter is also an element of Space.

The outer Space is the emptiness between all the celestial and terrestrial bodies, including our planet Earth. Scientists call it a *void* or a *vacuum* that consists of particles that are very far apart. These particles include *electromagnetic radiation, magnetic fields, plasmas, Space dust,* and *cosmic rays.*

But more importantly, the outer Space is the gap between our blue planet and the things we see beyond the sky. It is the distance between our planet and the Sun, the Moon, the Stars, and everything else you can see out there. Space is the blackness that you see when you look up at night.

But why is the outer Space black at night? The reason is the same when you turn off the lights and your room suddenly appears black. Nothing out there is visible without light either, and the outer Space itself emits no light. But there are a lot of things in outer Space that either generates, emits, or just reflects light from other sources.

And in Space, there are probably more things that give off generated light than we can ever count. There are lots of Comets, Asteroids, and other Planets that are visible in the blackness of Space, and they do not generate light. The reason we can see them is because these objects reflect light. Once that happens we can see them.

Space also has no friction to restrict speed and movement. This makes it possible for Stars, Planets, moons, and Space rocks to move freely.

However, a lot of these elementary bodies forcefully follow an *orbit*.

This is the force of what we know as *Gravity*. Gravitational pull is theorized to be the result of the weight of a Star that bends the fabric of black Space or dark matter, and causes nearby bodies to remain in that area where the Space is bent or sunk in.

Also thanks to gravity, there is something to prevent you from *falling into the sky.*

Just imagine how scary it would be to float into Space. If there were no gravity then there would be no weight nor would anything stay where it is, even you would not be able to stay on the ground.

These are just some of the things we know. There are still many things we do not know about outer Space. But we are still learning a lot; thanks to sciences like "Astronomy". Astronomy is the study of Stars Planets and Nebulae. It is known among all the other sciences as the study that helps the science community better understand what elements are out there in outer Space. This science is given the name Astronomy because starting with the root word "Astro" which is a prefix meaning "Star" or "Celestial Body", the word suggests what it studies. There are other sciences that use that root word as well but we will be looking at the one that goes with this book, and that is "Astronomy".

2 ASTRONOMY

The word Astronomy as mentioned in the last chapter comes from the word "Astro" which is derived from the Greek words *Astron* which also means *Star* and *Nomia* which means *law/culture*. Aside from what we have learned that Astronomy is the study of the Stars, Planets and Nebulae, it is ultimately the study of anything found in Space, because as man progressed in science more was discovered in Space to this day than those three main descriptive elements.

Essentially, everything else that you will learn from this book is under the science of Astronomy. But first, how did Astronomy or the study of everything in outer Space even begin?

One can say that Astronomy began when men first laid eyes and tried to interpret the *visible Space* in the night sky. At first, their observations were limited to how the positions of things change in Space. Particularly these Stars visible from Earth formed patterns that are known as constellations, But despite this, Astronomy back then still contributed a lot on what we know today. How did we know that one year equals 365 days? How did we know if its spring, summer, autumn, or winter?

How did we come up with the yearly calendar? Astronomy contributed to answering these questions in one way or another.

When we developed as a civilization, we grew more and more curious about outer Space. There were astronomical *observatories* constructed in civilizations like *Greece, China, Mesopotamia, Egypt,* and *Central America* that are meant to observe the visible Space. These Observatories are equipped with ground based telescopes, but in early times, Astronomy was mostly about *mapping* the positions of celestial bodies, including Stars and other Planets.

There were many key inventions that forever changed the world of Astronomy. One of the most important invention is the *Telescope,* which was coined by *Galileo Galilei* after improving upon the work of some historical scientists like *Hans Lippershey, Jacob Metius* and *Zacharias Janssen.* Without the telescope, Astronomy would never have evolved into a reliable form of science. It would have been just a sort of anyone's guess as to what is out there. But there is just no way we could have come this far without someone inventing something we could use to see far away from Earth, right?

With modern technology, Astronomy has reached new heights. It is also a science that is a cousin to Astrophysics. A word that also uses the root word "Astro" meaning "Star", only this time the root word is used with the word "Physics".

Not only are we adept in optical Astronomy or the Astronomy involving the visible Space, we can also study the outer Space using all kinds of radiations; including infrared, ultraviolet, gamma-ray, and X-ray.

So scientists that study physics can do that using the astro plane or Space. It is an awesome thing to be able to see far away and observe the movement of a planet and its satellite or even a Galaxy and its center, but even better to be able to scientifically identify the compounds and their physical   properties,   to   know   the   elements.

3 STARS

With early optical Astronomy, it was once believed that the Sun was larger than the Stars. It was also believed that the Sun and moon used to revolve around the Earth. When we started to learn more about the *Stars*, everything changed.

So, what exactly are Stars? Throughout history, Stars were interpreted in many ways. For a long time, Stars were believed to be balls of super-hot

gases. Since most Stars are massive in size, they can produce enough heat to turn its atmosphere into plasma. With the exception of *brown dwarf Stars*, Stars are technically huge balls of plasma. Well actually 90 percent or our closest Star the Sun, is plasma.

Plasma is a mixture of electrons and protons that were striped from the hydrogen atoms; as they hit onto one another they produce an awesome *nuclear fusion reaction* that releases energy. As hot gasses rise from the energy, they cool and fall back onto other gases that are rising and that just keeps going causing the movement of the Stars surface to happen. There is a whole lot of technical information of what is actually making a Star be what it is but that is a more advanced study than the scope of this book.

One of the important things we found out is that there are different types of Stars. These Stars vary greatly in mass, size, appearance, and *age*. There are the *main sequence Stars, red giant Stars, white dwarf Stars, Neutron Stars, variable Stars* and *brown dwarf Stars*.

As far as the existence cycle goes, Stars can exist prominently for billions of years. The oldest Star known to this day is the *HD 140283* or the *Methuselah Star,* which is estimated to be around 14.46 *billion Earth* years old.

The main sequence Stars are Stars that are in its *healthy* life. **(It is important to note that scientists say "life" when talking about Stars however a Star is not a living thing like we are. A Star is an element compound and is not alive!)**

But once their *hydrogen fuel* supply runs out, they expand into *red giant Stars*. If a Star is roughly the same size as our Sun, it will turn into a *white dwarf Star* and simply fade away after a very long time. If a dying Star is large enough, a *Supernova* would occur and leave behind *Neutron Stars*. The death of very massive Stars will also cause *Black Holes; Supernova's* and Black Holes will be discussed in later chapters of this book.

As compared to the Earth, Stars are *giants*. The Sun being the nearest Star is still *92,965,050 miles* or *149,598,261 kilometers* away from the Earth.

And although the Sun may look very large when we look at it from the ground, it actually belongs to a group of smaller Stars known as *yellow dwarfs*. How tiny are we, huh? ☺

But the word "dwarf" may actually be an understatement where it comes to these Stars and their hugeness. Our Sun, for example, is *1,412,000,000,000,000,000 cubic kilometers large,* (fifteen zeros) with a radius of *432,376* miles or one hundred times that of the Earth. That means the Sun can hold around *one million* Planet Earths inside it!

Considering the size of other Stars, the Sun is still extremely small. Or are the other Stars extremely large? The largest Star ever known in Astronomy as of this day is the red hyper giant *VY Canis Majoris*. Its estimated radius is around 1800-2100 *times* the radius of the Sun. But when you look at how much Space is out there and just how much room there is available from one object to another then it is understandable that there will be many elements in Space that are just so large our little brains cannot possibly imagine its size.

Or rather just think of it like this; if you were to go there to that Star that is eighteen hundred times larger than the Sun, in a Space ship, you would stay back a certain distance so you don't burn up. And that would make the Star appear to be about the right size. The Stars of the universe are the most abundant element out there. In Space there are more Stars than anything else, in fact, you might even say that it is the Stars that have ultimately caused all the other elements found in Space to exist. They cause gravity, they explode and new Stars form out of the remnant, they orbit and cause galaxies to form by sticking together and so on and so on…the Stars in   Space   are   the   Stars   of   Space.

4 PLANETS

If you look at the big picture, you can say that you yourself are technically in Space. A planet, including our planet Earth, is also an astronomical body that is found in Space. Although Planets are incredibly small compared to Stars, a planet should be large enough that its shape is rounded by its own gravity. This is because its *gravitational force* is originating from the very center, the planet's core, – pulling and holding everything together, forcing the planet's crust and other surface material to stay down.

So where do Planets come from? Astronomers believe that Planets come from the fragments of a newly-born Star. This occurs in the process known as *accretion*. These fragments stick together into large masses and get caught in the Star's gravity. If they move fast enough, these fragments will be trapped in an *orbit* due to *Celestial Mechanics*, the study of orbital dynamics

that are under the influence of gravity, with both astronomical bodies and Spacecraft produced by man.

In time, these fragments continue to grow in size by colliding and sticking to other fragments. This is how Planets, Moons, Asteroids, and Comets are formed.

All these astronomical bodies make up what is known as a *Solar system or Star system.* Of course, in time we have discovered that there are countless of other Planets orbiting countless of other Stars. All of which are caught in that original gravity that created them; their host Star. Now I don't know about you but that development of a Star system is just mind bending! Ah, but it is also Space, or Dark Matter bending.

In our *Solar System,* there are *8* Planets orbiting the Sun, and we have named them: *Mercury, Venus, Earth, Mars, Jupiter, Saturn, Uranus,* and *Neptune.* Mercury is the planet closest to the Sun. It is one of the *Terrestrial Planets* in the Solar System – meaning it is composed of rock and metal and has terrain.

It is also the smallest planet in our Solar System known to date. Since it has no atmosphere, there is nothing to retain the heat coming from the Sun. This is why Mercury's temperature ranges from extreme heat to extreme cold.

Venus, the second planet closest to the Sun, is another terrestrial planet. It is also regarded as the Earth's "sister planet" because they are similar in size, mass, and are quite close to each other. Venus is known as the *hottest planet* in our Solar System because its atmosphere is composed mostly of *carbon dioxide* – trapping heat as it enters. Venus can be seen from Earth as the second brightest celestial body at night; next to the Moon.

The third planet closest to the Sun is our blue planet *Earth*. Everybody knows that our planet is special because it is the only one in our Solar System that is known to support life. Its surface is covered mostly by water. Earth is the largest terrestrial planet in the Solar System. It is also known to be the  densest planet. Although modern technology has discovered other Planets similar to Earth, our planet remains the only one known to host life.

Mars is the fourth planet closest to the Sun and is the last of the 4 terrestrial Planets in the Solar System. It is also called the "Red Planet" because of its reddish color. This is because Mars's surface is rich in *iron oxide* or *rust*. It is now commonly believed that the red planet once contained water. This is due to the water molecules and ice water found in returned samples. The planet is also filled with geological features that support the speculation that bodies of water once existed on the planet's surface.

Jupiter is the fifth planet closest to the Sun and is also known as the largest planet in the Solar System. In fact, it is bigger than all of the other Planets combined. Jupiter is composed mainly of *helium* and *hydrogen*, which is why it is often referred to as a gas giant. A popular characteristic of Jupiter is its Great Red Spot.

This spot is a super-storm that is three times the size of the Earth.

Saturn is the sixth planet closest to the Sun. It also ranks next to Jupiter as the second largest planet in the Solar System. Just like Jupiter,  Saturn is also a gas giant. It is unique in our Solar System for its ring system. The rings are composed mainly of ice particles and some rocks and dust which are caught in orbit based on Saturn's gravity.

The next planet that shows up after Saturn is *Uranus*. Just like the gas giants, Uranus' atmosphere is composed mainly of hydrogen and helium. However, it contains far more *water, methane, ammonia,* as well as other *hydrocarbons* – also called ices. This is why Uranus is called an *ice giant.* Uranus has the coldest atmosphere in our Solar System.

 The last and farthest planet from the Sun is *Neptune.* Just like Uranus, Neptune is an ice giant. But unlike Uranus, Neptune's weather patterns can be visibly observed thanks to its atmosphere. The interior of the planet mainly consists of ice and rock.

Interestingly, Neptune also has a ring system like Saturn. But Neptune's ring system is faint and far less dense than Saturn's.

In the past, there were *9* Planets including *Pluto*. But with the discovery of the *Kuiper Belt* (see *Chapter 7*), the body Pluto was no longer recognized as a planet. Instead, it is now only considered as one of the debris found in the Kuiper Belt.

There were also other celestial bodies that may be considered to be Planets. But, just like Pluto, they were instead recognized only as *dwarf Planets*.

Like regular Planets, dwarf Planets orbit around the Sun. In our Solar System, there could be hundreds or thousands of dwarf Planets to be formed. But today, only *5* are officially considered: *Pluto, Haumea, Makemake, Ceres,* and *Eris.* So all things found in Space and studied by man have been done from a planet that is in a Star system and is the third planet from its Star and the Star is in a Galaxy.

The planet is Earth. The Galaxy is the Milky Way.

5 NEBULAE

Outer Space is filled with beautiful things like Stars and Planets. You can tell this just by looking at the night sky. But one of the most amazing things out there to behold is the *Nebulae*.

All of the beautiful things in Space have amazing stories that they can tell. A Nebula is no exception. They teach us a lot about how things begin or come to be and end or cease to exist in outer Space.

Just like there are clouds in the sky, there are Nebulae in Space. Nebulae are massive clouds of hydrogen, helium, dust, and other ionized gases. They are formed when astronomical materials, mostly gas, collapse under their own weight. The center of a Nebula may be the birthplace of a very big Star.

Nebulae may also be the result of an exploded Star.

Whenever a huge Star reaches an end and cannot burn bright any longer, its outer layers get released into Space. In many cases when this happens with very big Stars they experience what is called a *Supernova*.

While there are countless Nebulae waiting to be discovered in deep Space, there are certain famous Nebulae known to us right now. One of which is the *Crab* Nebula, which is the result of a Supernova (a Star that exploded). At the center of the crab Nebula is a Pulsar. A Pulsar is a spinning Neutron Star that lights up the center of the Nebula usually blue to blue-green in color.

The Neutron Star is so tiny that it is only about 12 kilometers or 7 miles in diameter.

This type of Star packs dense mass of neutrons and it is what is left of a Star after it explodes. The Nebula is also known as the remnant of the explosion of a Star. This is so fascinating that we can look in Space through the telescope and see the beauty that is left behind from a Supernova and the wonderful awe that is in its remnant. Just like on Earth when you see beauty in nature; but when you look in Space you think of blackness of night only to find the awesome beauty the elements of the universe have to show!

It is known that if two Neutron Stars should collide they create gamma rays. It is also theorized that if the Star which has exploded were more massive, then it would not have created Nebulae with a Pulsar or spinning Neutron Star, but a Black Hole would be the result. (We will go over Black Holes later in this book.)

Another very interesting Nebula is the Eagle Nebulas' *Pillars of Creation*. They are called that because they look like three magnificent pillars 7,000 light years away. When discovered, the Nebula was in the process of forming new Stars. However, it is now believed that this Nebula was already destroyed in a Supernova more than 6,000 years ago.

But since it is 7,000 light years away; its destruction is still not visible here on Earth. In other words, the light still needs to travel for 1,000 or more light years before it can reach us. What was not discussed yet because it is a bit more advanced than the scope of this book; is what a light year is.

In definition, a light year is the distance that light travels over the course of one Earth year, which is the amount of time that it takes our planet Earth to go around the Sun. So if something takes 2 light years to get here, then that means the object has to go 5,878,499,810,000 miles or 9,460,528,400,000 kilometers times two. In other words one light year is a long way!

The reason it is so long is that light is so fast, it goes over 186,000 miles or 300,000 kilometers per second so think about how far it would go in not a second or a minute or hour but a year!

Mind blowing stuff to understand the size of Space and how far away things really are, which is why this book is called "That Awesome Place called Space". Now back to the observations.

Another type of Nebula is the *Planetary Nebula*. Planetary Nebulae are formed when a Star reaches an age that it begins to have less fusion from its core or just gets too old to continue to burn and shine as a normal Star would. When this happens the gravity of the material at the outer part of the Star begins to take a toll on the Star itself and forces the inner parts to condense and heat up.

Then what happens over the next several thousand years is the heat from the center drives the outer part of the Star away. This creates a stellar wind that makes the Nebula look the way that it does and after that process is complete the remaining core elements are uncovered and heats the now distant gasses causing them to glow.

But being called Planetary Nebulae has nothing to do with Planets and some scientists question this name but it was thought that the Sun would one day create a remnant of this sort and that our Sun hosts Planets giving it the PN label. It is estimated however that although the Planetary Nebulae are common in the Milky Way Galaxy with a number of around 10,000 PN's, that the Sun is not heavy enough to form a PN. The theory is that the Star must be at around 20% heavier than our Sun with a stellar existence cycle of only 25,000 years.

So that means these Planetary Nebulae are formed by only short lived Stars.

6 GALAXIES

If you think Stars, Nebulae, and Solar Systems are big, think again. *Galaxies* are made up of *billions* of Stars, Planets, Nebulae, and a lot of other celestial bodies. For example, our Solar System is inside the *Milky Way* Galaxy, which is known to have around *300 billion Stars*.

Take a look at the center of the Milky Way from Earth in the picture.

We have not discussed constellations of Stars as of the writing in this book because it is viewed as a different subject and perhaps an advanced side-track of the teaching of this material. But if you learned anything about constellations in school then you would have heard of the constellation Sagittarius. In the picture of the Milky Way Galaxy taken from Earth you will see the center of the Galaxy shining bright if you telescope out from the southeast part of the sky during late summer, just behind the constellation Sagittarius. (Now this may require an extremely powerful telescope.) But there in the night sky is the center of the Galaxy from which this Planet Earth resides.

It is simply amazing because we can see all kinds of Galaxies in outer Space but we cannot really see the Milky Way from a full view like the others. That is because we are in the Milky Way Galaxy.

It would take a Space ship and someone on board to go out of our Galaxy and go many light years away and then turn back and use a powerful telescope from the Space Ship to see the Milky Way Galaxy in a full view. So unless we get in a Space Ship and go far far away in Space we will never actually have a picture of the Milky Way in full view like we do of so many other Galaxies, but hey we can always see the Milky Way from the inside! ☺

So what exactly makes a Galaxy?

Every Star in a Galaxy is held together by its own gravity. Of course, a lot of these Stars, including our Sun, have their own Planets. A Galaxy like ours has a *spiral shape* that's illuminated by the billions of Stars it is made up of. But an egg-shaped Galaxy is even more common.

Our Galaxy, the Milky Way, is incredibly large. It has a diameter of *100,000 light years,* meaning it will take 100,000 *years* to go across if you're moving at the speed of light. In comparison, it takes only 8 minutes for the Sun's light to reach us here on Earth.

And remember that the speed of light is *186,000 miles per second.* Just imagine how small our Solar System is compared to our Galaxy!

But no matter how big our Galaxy is, it is only considered as a *mid-range spiral* Galaxy. The largest galaxies are the *elliptical galaxies* that can contain trillions of Stars!

The largest Galaxy we know as of now is the *IC 1101*. It is a supergiant elliptical Galaxy located at roughly a billion light years away from Earth.

How big is the IC 1101 Galaxy? Only *6,000,000 light years* in diameter! That is more than *50* times the size of the Milky Way!

But if there are giant galaxies, there are also the smaller, *dwarf* galaxies. Dwarf galaxies are the smallest kind of galaxies. Most of them are only around *200 light years* in diameter. They also have far fewer Stars than mid-range spiral galaxies. That is of course because smaller means fewer. All the bodies of mass within a Galaxy are contained by the gravitational pull of the center, but a Galaxy moves so slightly, or slowly, that it takes literally millions of years to notice any difference and of course we don't live that long to actually see that happen.

No one can really accurately say just how many Galaxies there are in the universe because it is just not something we as human beings can really count. Sure we can see out into Space and Astronomers have been tracking and naming the elements of what they find out there, since the earliest centuries of mankind. They also do their best to map what they know about the universe and where it all is located physically.

That mapping is done much better now that NASA has put a large telescope in Space, which means it is outside the Earth's atmosphere and avoids having to look out into the universe through the haze of that atmosphere. This telescope is known as the "Hubble Space Telescope".

A large 44 foot giant telescope that orbits the Earth and is controlled by Scientists who can now see farther out into Space than ever before in the history of the study of Space. So by using the Hubble they can see a Galaxy back in time. This is because the Space Telescope will find light that has not made it to Earth yet, but if that light did make it to Earth then the Galaxy that put off the light my look different.

In fact, most of the light in deep Space seen from the Space telescope has yet to make it to Earth, so scientists can actually see the movement of something in Space like a Galaxy but it is with the help of a device that can go back into time. *Imagine this*: If you are around in a billion years then the night sky will look a lot different because there will be all kinds of new elements of light that made it to Earth.

So how many Galaxies are there in the universe? In the last page it was said that no one can really tell just how many Galaxies there are in the universe but there is what is known as an *observable universe*. Scientists have counted the galaxies in the observable universe by using the Hubble Space Telescope.

They Started by counting the galaxies in one region and then multiplying that number up, based on theory and fact, to come up with a number of

100 Billion galaxies. But those one hundred billion galaxies are estimated and that estimate is talking about the observable universe.

If you gaze into a photo taken by Hubble you can see that the universe is there even beyond those galaxies. That is to say; look very closely in between the galaxies shown and you see that the blackness of Space keeps going even farther and farther behind the galaxies, so even Hubble cannot see all of what is there because it can only see so far.

In fact it is in theory that the universe is basically infinite. If you are not familiar with that word, it means never ending. We exist in a universe that has no end. There is no wall out there marking the end of the Space. So when giant telescopes get better, in the future we should be able to see two hundred billion galaxies. But it is things like wanting to know where the universe ends that drives young and old minds alike, to try and find out more about the theories of the infinite universe.

Maybe the answer is at the Black Hole which is talked about later in this book. Could there be another place that is also called a universe at the other side of going through a Black Hole?

The world's most brilliant scientists can formulate using mathematics and the periodic table of elements as well as known data of all that exists out there and come up with some possible answers or explanations, but Space will always be home to what man just does not know. We may never know just how many galaxies exist, but we do know they are out there beyond the Earth's sky. That is why the "Big Bang Theory" is theory.

Theory = Educated Guess!

So we learn, we equate, but then we can only guess based on the equation hoping that the theory is accurate. And often times we find that the theory of a given wonder that was once held, based on new fact must be debunked. However sometimes an educated guess is all we got!

7 ASTEROIDS, METEORS AND COMETS

We talked a lot about the giants of outer Space, so let's take a look at the smaller celestial objects. Outer Space is filled with massive things such as Stars, Nebulae, and Planets. But in Space, there are also some small bodies. And with Astronomy, the words small bodies basically mean rocks, and in Space, there are a lot of rocks; *Space rocks*.

There are three main classifications of Space rocks: *Comets, Meteoroids,* and *Asteroids*. It is easy to confuse them from one another. But if you look closely, there are many differences between these three types of Space rocks.

First we look at the Comet. It is not really made entirely of a rock; rather it is composed of dust, gas, ice, and small particles of rocks.

Comets are also sometimes referred to as the "Dirty Snowballs" of Space. This is because they contain ice as one of the primary materials.

Comets are quite beautiful themselves as they are usually observed with a tail. But these tails actually only appear when a comet is close to the Sun. As Comets approach the area of the Sun, its ice particles and some gases start *evaporating*. This process forms a *cloud* around the *nucleus* or the center of the Comet. This cloud, also known as *Coma,* is pushed away into outer Space by *Solar Winds*. That is what forms the Comet's tail. Although an average Comet's nucleus is only around 1 to 10 kilometers wide, its tail can reach *one hundred million miles* in length!

The next Space rock has three potential names: *Meteoroid, Meteor,* and *Meteorite*. But what is the difference between them?

First of all, this type of Space rock is very small. They are usually only *10 meters* or *less* across. A *Meteoroid* is actually just a fragment of another Space rock such as a Comet or an Asteroid. Some Meteoroids that are close to Earth get pulled in by its gravity. And as they enter the Earth's atmosphere, they start burning off. When this happens, they are called *Meteors*.

Whenever you wish upon a shooting Star, you are wishing upon a *Meteor*. This is because shooting Stars are technically meteors that are large enough to be seen from a distance. Meteors are actually very common, but only a small percentage will ever hit the surface of the Earth.

When a Meteoroid survives the Earth's atmosphere and actually strikes the Earth, it is then classified as a *Meteorite*. Meteorites vary greatly in size and mass. Some can be as small as a grain of sand while others can be as big as a football field. It is also believed that a very large Meteorite that created the *Chicxulub Crater* in Mexico resulted in the extinction of the Dinosaurs.

Another main classification of Space rock is the *Asteroid*. These astronomical bodies are made up of rocks and sometimes metals like iron and nickel. They can be as small as a few meters wide to as large as an entire state. Larger asteroids are sometimes referred to as *Planetoids*. Some also believe that asteroids are Planets that failed to form.

Some of the larger asteroids were found with a Satellite or Moon of their own. This is a smaller body that orbits the Asteroid just like with Planets. These Asteroids reveal that there is some relevance in the behavior of all the celestial and terrestrial bodies discovered in outer Space. They show that even a Space rock can be a lot like a Planet and Moon even though it is not round. But there are however huge differences in that Stars and Planets are formed round for a reason based on the matter they are formed of and the original dynamics of their formation.

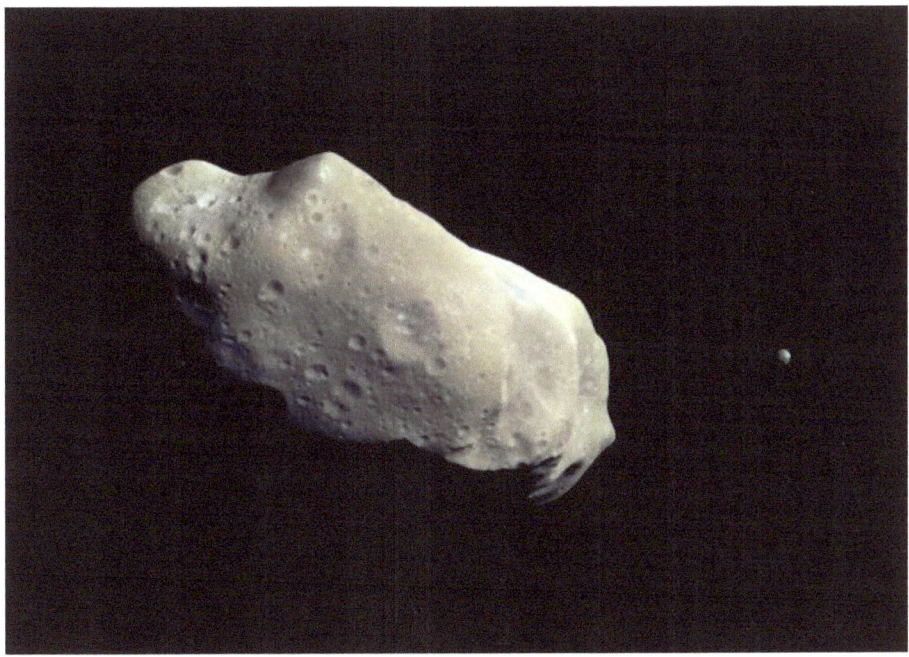

Most of the Asteroids in our Solar System are grouped in what is known as an *"Asteroid Belt"*. The Asteroid Belt is located between the orbits of Mars and Jupiter. This is also referred to as the "Main Belt". Other than this, there are three other Asteroid groups in our Solar System: These are the *Trojans,* the *Scattered Disc,* and the *Kuiper Belt.*

Asteroids and other Space Rocks that are colliding with each other cause huge explosions. Meteorites that hit Earth, like the *Chelyabinsk Meteor,* may also cause very devastating explosions. But in Space, no other explosion is as powerful as a Supernova.

Another very large Comet known as 67P/Churyumov-Gerasimenko got a lot of attention when the European Space Agency sent a Spacecraft called the Rosetta to the 2.5 mile or 4 kilometer sized Rock to explore it up close. Once there they actually put a Lander on the Comet, which is quite an amazing accomplishment, just like NASA putting a Lander on the Planet Mars.

The Comet is around three hundred and seventeen million miles from Earth or 510 million kilometers away. As time goes on we continue to advance with putting Spacecraft out on missions and soon you may even start seeing manned missions. But in this case it looks like the odd shaped Comet was a terrestrial body of opportunity for Scientists.

8 SUPERNOVAS

When massive Stars reach the end of their cycle, they do so spectacularly. It results in a magnificent explosion that is also considered as one of the most powerful events that happens in outer Space – The Supernova.

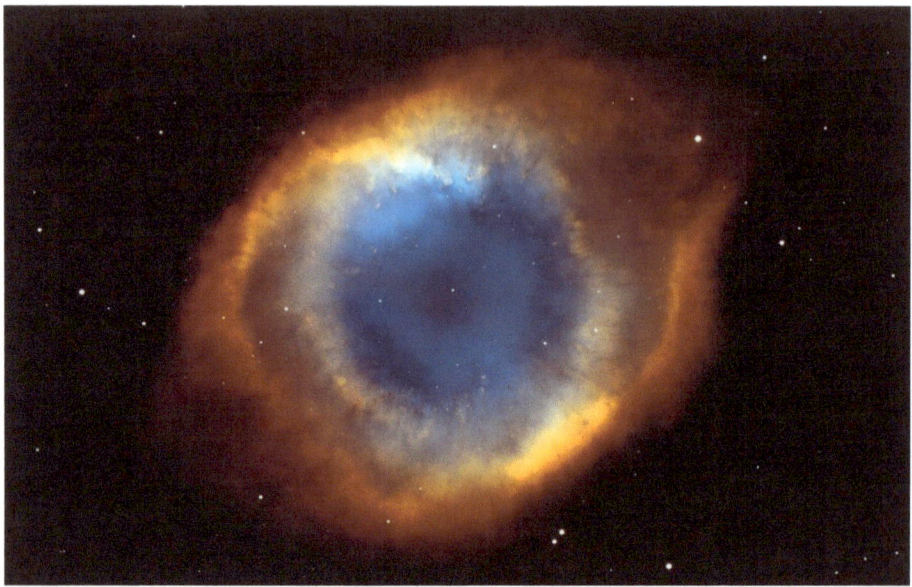

A Supernova can be the result of the end of existence of a Star. Just like cars, and many other machines here on the Earth, that use fuel to run, Stars use fuel to shine out there in Space. But since Stars are a little more complicated than machines not all of them will simply stop when their fuel runs out.

Not all Stars will produce a Supernova when their cycles end either. For example, our Sun is not what we call a massive Star and in theory will only expand into a red giant when its hydrogen fuel runs out.

Although there are billions of Stars in our Galaxy, Stars that end with Supernovas are rare and since they produced the Supernova it is not really the end of that astronomical element because now it has formed a remnant which often will have a Pulsar; and thus the cycle starts over. It is predicted that only several large Stars in our Galaxy will cause Supernovas within the next thousands to millions of years.

These Stars include *Antares, Betelgeuse, Eta Carinae, RS Ophiuchi, Rho Cassiopeiae, U Scorpii,* and *VY Canis Majoris.* In our Galaxy, the last recorded Supernova is discovered by *Johannes Kepler* in 1604. It is known as type la Supernovae because the remnant that Kepler saw was a white dwarf Star that was known to be much heavier in its elements than our Sun and it exploded. So it shows that the idea that only massive Stars will have a Supernova is not always the case.

When a big Star like the ones mentioned above runs out of fuel, its mass squeezes into its core. When this happens, its core becomes so heavy that it collapses from within its own gravity. This results in a giant explosion that will be brighter than an entire Galaxy for a brief moment.

The Supernova will rapidly release what remains of the Star's composition into deep Space. Eventually, those remains will be reused to form other Stars, Planets, and other celestial bodies.

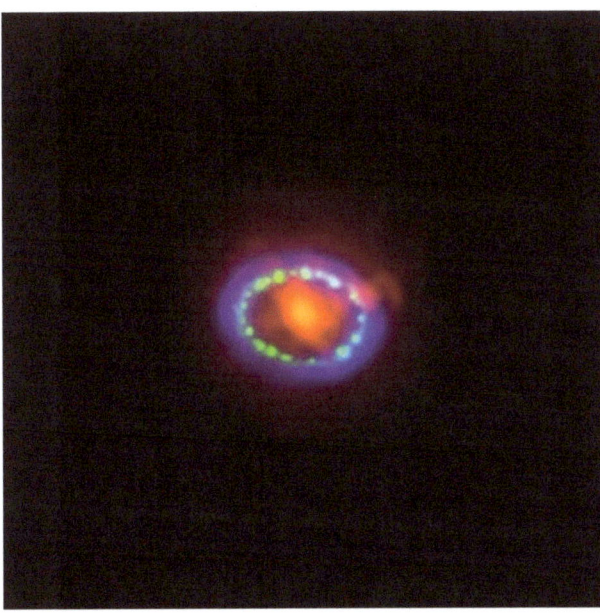

Captured here you can get a look at what a Supernova actually looks like when zoomed in on from a telescope. But if you zoom out this will appear as a bright reddish circle among all the other Stars and Space elements that are around it.

Although it is said that the Supernova is just not that common in the Milky Way Galaxy, we need to take in account that there are billions of Galaxies to consider when looking for Supernovae. In fact there are Stars scattered out among the Universe that happen to not even be in a Galaxy.

There are Stars existing in Space in between Galaxies and some of these Stars are seen to be part of binary systems, which means they gravitate around one another forming an orbit all of their own. But it could be that any of the Stars there are to be found, may potentially end in a Supernova causing this event to be more of a common part of the overall Universe.

Here is another look at a Supernova. This one has a great proportion of the explosion emitting what we call *"Cosmic Rays"* These are found throughout Space and are of very high energy portions and atomic nuclei. Other places astronomers find Cosmic Rays are at the center of galaxies that have active atomic nuclei. So as Scientists study deep Space they always find more and more evidence of Stars that went Supernova, if not even catch one in action.

There is also another way for a Supernova to occur. When two Stars collide, what else can you expect other than a gigantic explosion right? In what is known as a *Binary Star System,* like mentioned already, two Stars orbit at a similar point. If they get close enough to each other, they will eventually merge into a bigger Star. This will also eventually result in a Supernova explosion.

So we have discussed the Supernova in general that leaves a beautiful remnant and sometimes even a pulsar or Neutron Star remains spinning in the center. We have talked about our own Galaxy and how common Supernovae are, the white dwarf and the Binary Stars all of which causes the Supernova to take place.

But sometimes, a powerful Supernova can also lead to something even more destructive, and that something is: *"The Black Hole".* So next let's talk about   the   Black   Hole   and   learn   what   they   are

9 BLACK HOLES

When a massive Star that is at least 10 to 15 times larger than our Sun explodes into a Supernova, it is believed that the result is a Black Hole. The core of the Star becomes a Black Hole.

Black Holes are among the most extraordinary phenomena known to Astronomers; and to the study of the entire *Universe*. They are extremely dense and massive; resulting in a very powerful gravitational pull that *even light cannot escape.*

This is the reason why Black Holes reflect no light at all. However, this powerful gravity only affects a particular region around the Black Hole. This region is called the *event horizon.*

The event horizon can be considered as a Black Hole's buffet table. It can swallow Planets, Stars, radiation, and even entire Solar Systems.

The more matter a Black Hole consumes, the heavier and more powerful it becomes. Scientists have yet to find any evidence of a Black Hole that would be close to our Solar System so we may be safe from falling victim of one however things are indeed unpredictable in outer Space.

But you shouldn't worry about a Black Hole swallowing Earth.

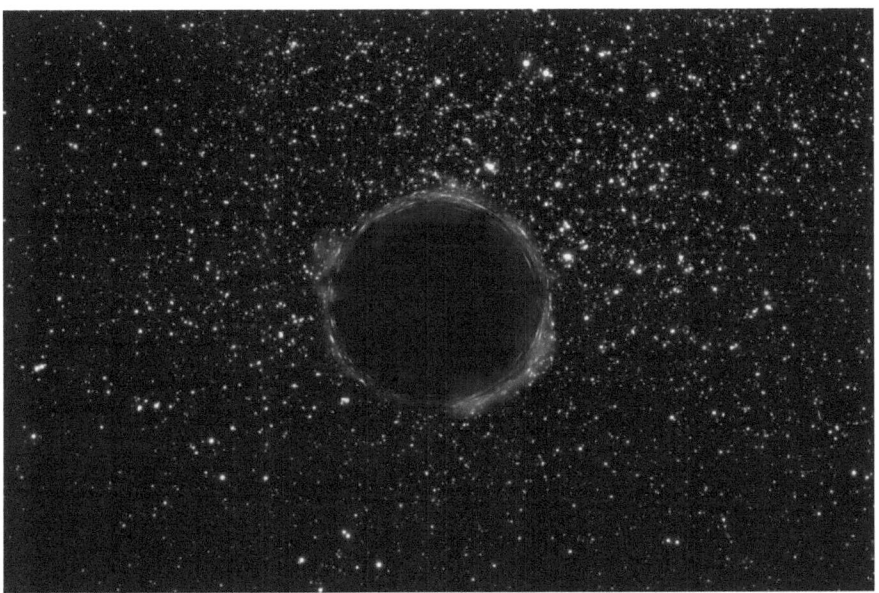

Black Holes are also believed to be different sizes. If a Black Hole becomes large enough, it turns into a *Super Massive Black Hole.*

Super massive Black Holes are Black Holes that are extremely massive. One super massive Black Hole can be the equivalent of more than *1 million suns.* Scientists also believe that a super massive Black Hole lies at the center of most large Galaxies. It is also believed that super massive Black Holes were created at the same time as the Galaxy where it is found.

In our Galaxy, Astronomers agree that a super massive Black Hole is probably resting beyond the constellation *Sagittarius,* found at the center of the Milky Way. It is said to have a mass that is equal to around *4 million* suns.

But if there are super massive Black Holes, there may also be Black Holes that are extremely tiny.

They are believed to be so small that they evaporate in an instant. At least, this is what Astronomers believe. As of today, Astronomers are still looking for more evidence for these "Mini Black Holes" or *Micro Black Holes* in Space.

But what exactly happens inside a Black Hole?
Unfortunately, no one in the world has a definitive answer. Not only are Black Holes hard to study, they are basically impossible to study *directly*.
One reason is because since light gets swallowed by it, you can't actually see a Black Hole. You can only observe all the other elements around it.

Is it a portal to another universe? Will time stop once you're inside? There are plenty of theories in existence to explain what happens in a Black Hole. But right now, no one can be sure. If we are going to be so amazed by the universe, all the elements of Space and outer Space, try to see as far into deep Space as we can, then we need to also be amazed at all the unexplained or unformulated answers as well. Meaning that if Mankind can explain what a Star, Planet or Nebula is made of, then that is great.

It is great that we know those things but there are always more things that we do not know as of yet, and may never know. Perhaps one day in the future we can send a probe into a Black Hole and find out what becomes of it. Maybe it will come out the other side of the Black Hole and see a different universe that exists parallel to this one and maybe there is a different Earth there that we don't come from but still has Dinosaurs on it.

It is the imagination of mankind that has us constantly looking out into Space and trying to figure out all the science and all the astrophysics as well as all the exact study of particles and atoms to find out what it all means and why it does what it does. And that is a healthy thing because it tells us all about ourselves. It shows us that Space is indeed an awesome place if we can stay out of harm's way! ☺

10 CONCLUSION

Thank you again for your purchase of this book!

I hope you enjoyed reading your book on Space! Maybe you have looked at magazines and other books on this subject before. I know there are a lot of books and other reading material available for understanding Space and the concepts of the Universe. In this book it was my goal to bring forth a simple outlook in what was hopefully the simplest terms I could use and still get the point across accurately. But I am glad to put it all together with some pictures as each chapter discusses its particular subject.

As in the front of this book I wrote how I always enjoyed the subject of Space and outer Space. I could gaze at Stars at night and ponder the mystery of all that exists out there for hours. So it is my hope that this book reaches those who are enthused to learn more like I was a child and still today, then maybe you will take it further and go into the field of Space study and Astronomy. But even if you don't, I hope the book can be a fun reference to look back at and share with others.

Finally, if you enjoyed this book, please take the time to share your thoughts and **post a review on Amazon**. It would be greatly appreciated!

Thank you!

Jon P Fox

ABOUT THE AUTHOR

Author: Jon P Fox

Born and raised in Michigan in 1962, Jon's grandparents migrated from Ireland and his parents were raised in East Detroit. Jon is a professional consultant and aspires in many different hobbies such as Art and Music along with Photography and Video. Jon likes to stay close to God by his attention to the existence of the Holy Spirit, and God's written word! You can find out the things he is very passionate about by taking notice to not only what he writes on his blog but the books that he writes.

Jon is the founder of up and coming ever developing innovation company JPF & Associates which is a small business that has published various paperback, Kindle, Nook, and Audio books and is working to grow this into much more.

Feel free to contact Jon at jonpf239@gmail.com Check out his Amazon profile here: https://www.amazon.com/author/jonpfox

Next Steps

-    Write me an honest review about the book – I truly value your opinion and thoughts and I will incorporate them into my next book, which is already underway.

Leave your review of my book here:

http://www.amazon.com/dp/B00VW3B77Y

Scroll down to the "Customers Reviews" section and click on the big yellow button that says: "Write a customer review"

Remember I value your thoughts and thank you very much!

Check Out My Other Books

Go ahead and click on the links (or type in the browser for hard copies) below, to check out the other great books I've published!

Making Time For God

http://www.amazon.com/dp/B00BGI458I

Making Time For God Volume 2

http://www.amazon.com/dp/B00F21LXF8

Making Time For God Volume 3

http://www.amazon.com/dp/B00TQ9QNM0

Cool is the Guitar

http://www.amazon.com/dp/B00KPG8CRM

You may also visit my author page here:

https://www.amazon.com/author/jonpfox

And my Facebook Page Here:

http://on.fb.me/1jqLYyC